Tex brought Max a box.

In the box was an ox.

"I brought you an ox!"

said Tex.

"I bought the ox!" Tex said.

"Yuck!" yelped Max.

"I wanted a fox!"

"A fox?" yelled Tex.

"I thought you wanted an ox!
You said ox," he said.

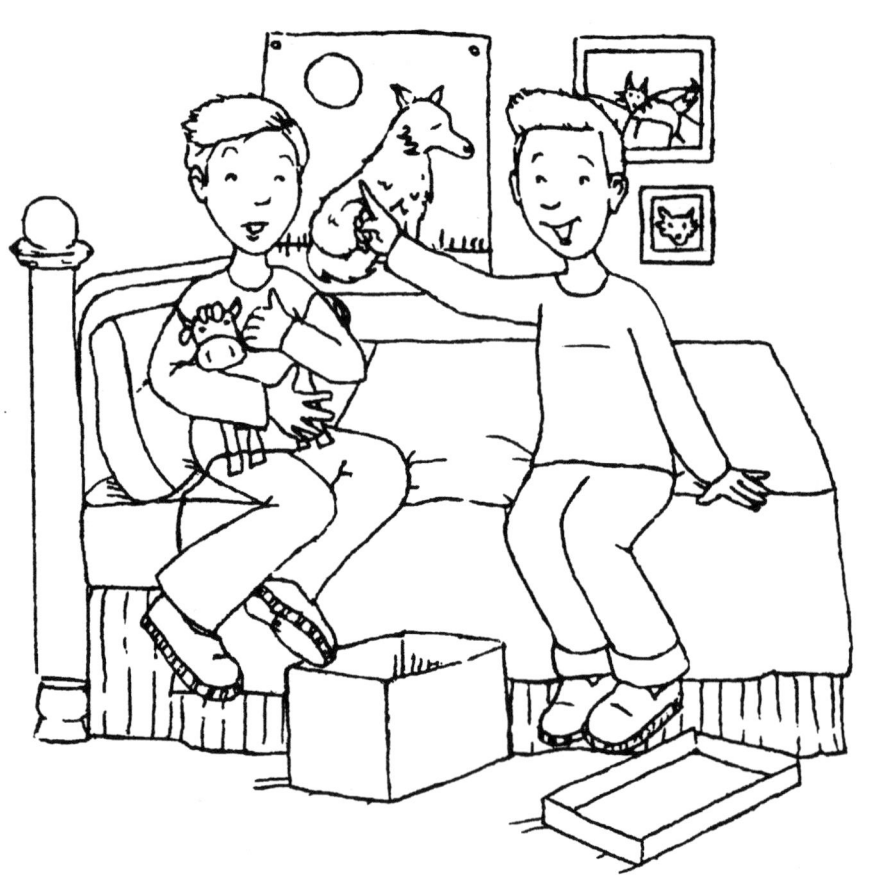

"I said fox, not ox," said Max.

"Ox, not fox!" said Tex.

So Tex and Max had a spat!
No one fought.
But Tex and Max did yelp
and yell!

"Max, Tex!" said Max's mom.

"Stop yelling.

You act as if you are two,

not six."

"Can't you fix this?"

Mom asked.

"Yes, we can," said Max.

"Yes, we can fix it," said Tex.

"Let's take it back," said Tex.

"And get a fox!" said Max.

And the ox went into the box!

The End